PID parameter tables for control of time delayed systems for minimized ITAE and IAE criterions

Roland Büchi

AF672087

Bibliografische Information der Deutschen Nationalbibliothek
Die Deutsche Nationalbibliothek verzeichnet diese Publikation in der Deutschen Nationalbibliografie; detaillierte bibliografische Daten sind im Internet über www.dnb.de abrufbar.

Impressum

Roland Büchi, 2022 ©

Herstellung und Verlag: BoD – Books on Demand, Norderstedt
ISBN: 978-3-7562-1789-2

Contents

PID parameter tables for control of time delayed systems for minimized ITAE and IAE criterions

PID controllers have been by far the most frequently used controllers for a long time. The control of time-delayed systems is challenging in particular. There are some heuristic methods known that can be used in the time and frequency domains. These are, for example, those of Ziegler Nichols, Latzel, or Chien Hrones and Reswick. The parameters that are found in this way result in asysmptotically stable systems. In most cases, however, they show transient behavior in practice, which still requires readjustment.
In this publication, parameter sets for PID controllers are presented, which can be used to control systems with time-delayed stable step responses. These minimize the quality criteria in the time domain, IAE, ITAE and ISE. Since the determination of the parameter sets is very computationally intensive, an approach from the field of artificial intelligence was chosen for their calculation.
The application of the parameter sets found is verified using an example of the control of a system from process engineering, which has a very large dead time. The parameter sets also take into account the controller output limitation that is relevant in practice. The parameter sets are scalable with the time constant and the static gain of the system. Therefore they can be used to control all systems with a time delay behavior.

1. Models for time delayed systems

Time-delayed systems demand special control requirements because delayed signals are very difficult to control. In practice, however, they are very common, especially in process engineering or in thermal systems, since the sensor often cannot be placed directly next to the actuator. In order to be able to classify such time-delayed systems in terms of control theory at all, a mathematical model is required first. There are many such models in the literature. They are used with delay elements or combined with several serial PT1 systems. They are discussed in the literature. However, the PID parameter tables presented in this document, refer to PTn systems with identical time constants. These are identical first-order systems connected in series, i.e. PT1 elements. These systems are also very common outside of process engineering and can be found in all engineering disciplines. The step responses of such elements result in delayed signals. In this way, the dead times can also be modeled well with linear models. Formula 1 shows the transfer function of the model.

$$\frac{K_S}{(s \cdot T_1 + 1)^n} \qquad (1)$$

Time- percentage method

A method that can be used well to model these PTn systems according to the above is the time-percentage characteristic value method. The step response is measured according to figure 1 and first, the times t10, t50 and t90 are determined.

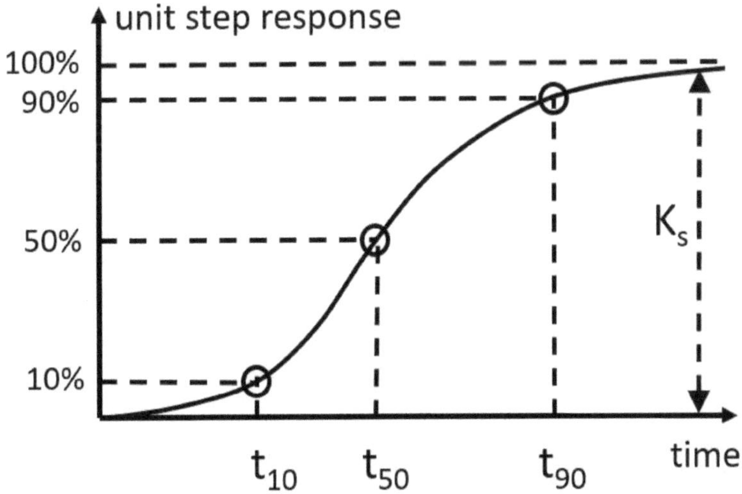

Figure 1: Step response of a time delayed system, time percentage method

Table 1 shows the parameters for the calculation of the time constant T1. The order n is determined via the parameter μ, according to formula 2 below.

$$\mu = \frac{t_{10}}{t_{90}} \quad (2)$$

Ks is generally determined with the formula:

$$K_s = \frac{static\ final\ value\ output\ after\ step -\ static\ final\ value\ output\ after\ step}{input\ value\ after\ step -\ input\ value\ before\ step} \quad (3)$$

And the identical time constant T1 of the n PT1 elements is calculated to

$$T_1 = \frac{\propto_{10} \cdot t_{10} + \propto_{50} \cdot t_{50} + \propto_{90} \cdot t_{90}}{3} \quad (4)$$

n, PTn	μ	α10	α50	α90
2, PT2	0.137	1.880	0.596	0.257
3, PT3	0.207	0.907	0.374	0.188
4, PT4	0.261	0.573	0.272	0.150
5, PT5	0.304	0.411	0.214	0.125
6, PT6	0.340	0.317	0.176	0.108
8, PT8	0.396	0.215	0.130	0.085
10, PT10	0.438	0.161	0.103	0.070

Table 1: Correlation of n, μ, α10, α50, α90

Method of turning tangent

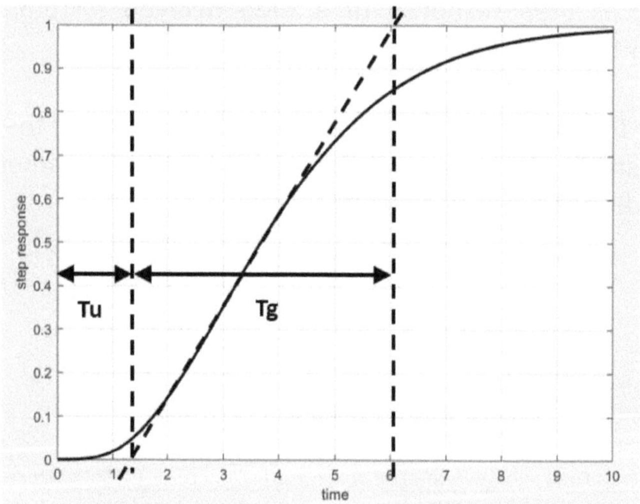

Figure 2: Step response to model a PTn- system, turning tangent, Tu und Tg

Another method for determining the n PT1 terms is the turning tangent method. In many cases one can simply measure the so-called delay time Tu and the rise time Tg of

the step response by placing a tangent at the point of inflection. From this one can identify the number n of PT1 elements connected in series and their identical time constants T1. It can be calculated after table 1, which is well known from the literature.
The parameters of table 2 are calculated using formulas 5 to 7 as a function of n.

$$\frac{T_g}{T_1} = \frac{(n-1)!}{(n-1)^{n-1}} \cdot e^{n-1} \quad (5)$$

$$\frac{T_u}{T_1} = n - 1 - \frac{(n-1)!}{(n-1)^{n-1}} \cdot \left[e^{n-1} - \sum_{m=0}^{n-1} \frac{(n-1)^m}{m!}\right] \quad (6)$$

$$\frac{T_g}{T_u} = \frac{\frac{(n-1)!}{(n-1)^{n-1}} \cdot e^{n-1}}{n-1-\frac{(n-1)!}{(n-1)^{n-1}} \cdot \left[e^{n-1}-\sum_{m=0}^{n-1} \frac{(n-1)^m}{m!}\right]} \quad (7)$$

n, PTn	Tg/T1	Tu/T1	Tg/Tu
2, PT2	2.72	0.28	9.65
3, PT3	3.69	0.81	4.59
4, PT4	4.46	1.43	3.13
5, PT5	5.12	2.10	2.44
6, PT6	5.70	2.81	2.03
8, PT8	6.71	4.31	1.56
10, PT10	7.59	5.87	1.29
12, PT12	8.38	7.48	1.12
16, PT16	9.76	10.78	0.91
20, PT20	10.97	14.18	0.77

Table 2: Correlation of Tg, Tu, T1 und PTn

2. Optimal parameters for PID controllers after minimizing the quality criteria IAE, ITAE and ISE

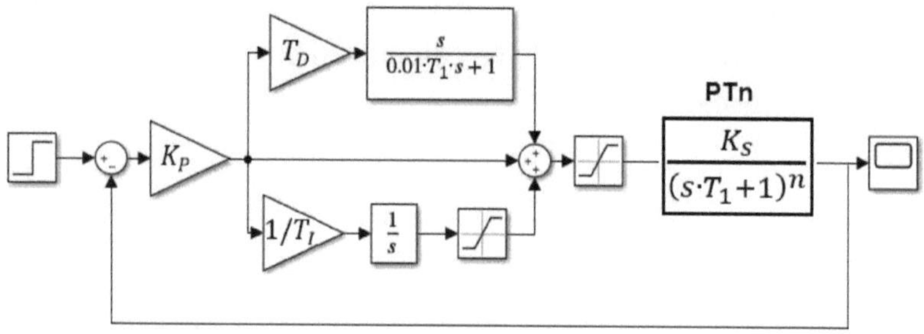

Figure 3: General block diagram of a PTn system controlled by a PID controller.

The block diagram of a closed loop system is shown in figure 3. The system to be controlled is the PTn found above with n PT1 elements connected in series. The parameters Ks, T1 and n found with the methods described above are also included. Various methods can now be used in practice to find the controller parameters of the PID controllers, Kp, Ti and Td. These can be model-based methods or those from heuristics. Heuristic methods are, for example, those of Ziegler Nichols or Chien Hrones Reswick. In control engineering, finding the parameters is a topic with great potential for optimization, because the parameters can be optimized using various methods in order to achieve good transient response. In addition to various methods from the frequency range, the criteria IAE, ITAE and ISE are also known. They relate to the time domain and describe the error area of a step response of the controlled system. These error areas are shown in figure 4.

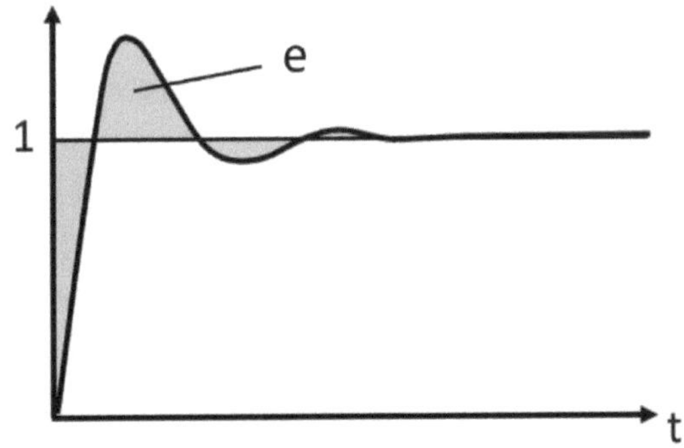

Figure 4: Error areas in the transient behavior of the closed-loop system according to figure 3, for calculating the IAE, ITAE and ISE criteria.

$$IAE\colon \int_0^{\infty} |e(t) - e(\infty)|\, dt;\quad ITAE\colon \int_0^{\infty} |e(t) - e(\infty)| \cdot t \cdot dt;\quad ISE\colon \int_0^{\infty} [e(t) - e(\infty)]^{2}\, dt$$

From this it can also be seen that the IAE criterion calculates the absolute value of the error area. The ITAE criterion additionally takes time into account. As a result, the error area is weighted more heavily as time progresses. The ISE criterion does not calculate the error area itself, but rather its square. This means that it does not have to be calculated with the absolute value, since the negative signs cancel each other out when squaring.

Because the time-delayed systems were approximated as PTn elements in Chapter 1, all PID controller parameters Kp, Ti and Td could also be calculated for all orders n and different Ks and T1. One would simply have to calculate the quality criteria for step responses described above. After that, the parameters for the smallest values of quality criteria could be presented as tabular values. The problem is that this has to be done for a multi-dimensional space (e.g. order n, Kp, Ti, Td, quality criteria, various limitations of the controller output). So that would take a very long time with the computers available today. This would have to be done with nested loops of all parameters. With the 6 parameters, this would result in complexity $f(x) = O(x^6)$. The problem would still be solvable in polynomial time, but the power of 6 is quite high.

Therefore, a method from the field of artificial intelligence according to Russell and Norvig was chosen, the hill climbing method. In this method, some of the parameters, in this case the parameters of the PID controller, are added to form a heuristic function. Then it is recalculated whether the quality criteria IAE, ITAE and ISE have become smaller. If so, then the new parameters are used as a new reference, otherwise the old ones are back calculated. In this way, and even after many iterations, the final values of the parameters remain at local minima of the quality criteria. The method requires much less computing time than a complete calculation in multi-dimensional space.

However, since it only finds local minima, many different random tuples of start values of the control parameters were simply used in practice when searching for parameters. Doing that, after the calculation many of the parameter solutions of the converged minimum quality criteria agree with each other. Therefore, one can assume with reasonably good certainty that the parameters found are actually the PID parameters Kp, Ti and Td, which either correspond to the absolute minimum of the quality criteria or at least come very close to them.

In control engineering, the use of this method of artificial intelligence is well chosen. It is often not possible to carry out complete calculations in the entire parameter space due to the temporal complexity of the calculation steps to be used. Since such methods always only calculate a part of the parameter space, the calculation time is greatly reduced and the results are parameter sets for excellent transient response. The cross-check to determine whether these are really the minimum parameters is carried out with several different start values for the parameters. Compared to the complete calculation, this still represents a significantly reduced computing time.

3. PID parameter table after minimal IAE, ITAE and ISE criteria

The table below contains the PID controller parameters calculated with Matlab / Simulink and the Hill Climbing method, which are based on the minimized quality criteria IAE, ITAE and ISE. The block diagram according to figure 3 serves as a basis. The controller output limitation is implemented on the one hand after the controller and on the other hand also after the integrator (anti windup) and is assumed to be +/- 2, +/- 3, +/- 5, +/- 10 in the calculations. Although the Anti Windup is never active here, it is inserted anyway, since in practice it can happen for various reasons that the controlled variable in the static end value does not reach the reference variable. It is particularly noteworthy that both the static gain Ks and the time constant T1 have a multiplicative effect on the controller parameters. This makes the table universally applicable for a very wide range of applications. The maximum parameter value for Kp·Ks is limited to 10.

	(Maximum controller output - controller output before the step) divided by (controller output for the stationary end value - controller output before the step)			
	+/- 2	+/- 3	+/- 5	+/- 10
PT1				
IAE	Kp·Ks = 10 Ti = 3.1·T1 Td = 0 (PI)	Kp·Ks = 10 Ti = 2·T1 Td = 0 (PI)	Kp·Ks = 10 Ti = 1.3·T1 Td = 0 (PI)	Kp·Ks = 10 Ti = 1·T1 Td = 0 (PI)
ITAE	Kp·Ks = 9.3 Ti = 2.9·T1 Td = 0 (PI)	Kp·Ks = 9.5 Ti = 1.9·T1 Td = 0 (PI)	Kp·Ks = 9.1 Ti = 1.2·T1 Td = 0 (PI)	Kp·Ks = 10 Ti = 1·T1 Td = 0 (PI)
ISE	Kp·Ks = 10 Ti = 2.7·T1 Td = 0 (PI)	Kp·Ks = 10 Ti = 1.6·T1 Td = 0 (PI)	Kp·Ks = 9.8 Ti = 1.5·T1 Td = 0 (PI)	Kp·Ks = 10 Ti = 0.2·T1 Td = 0 (PI)
PT2	**Tg/Tu:**	**9.65**		
IAE	Kp·Ks = 10 Ti = 9.6·T1 Td = 0.3·T1	Kp·Ks = 10 Ti = 7.3·T1 Td = 0.3·T1	Kp·Ks = 10 Ti = 5.6·T1 Td = 0.3·T1	Kp·Ks = 10 Ti = 3.7·T1 Td = 0.2·T1
ITAE	Kp·Ks = 10 Ti = 9.6·T1 Td = 0.3·T1	Kp·Ks = 10 Ti = 7.3·T1 Td = 0.3·T1	Kp·Ks = 9.6 Ti = 5.4·T1 Td = 0.3·T1	Kp·Ks = 9.8 Ti = 4.7·T1 Td = 0.3·T1
ISE	Kp·Ks = 10 Ti = 9.7·T1 Td = 0.2·T1	Kp·Ks = 10 Ti = 7.3·T1 Td = 0.2·T1	Kp·Ks = 10 Ti = 5.1·T1 Td = 0.2·T1	Kp·Ks = 10 Ti = 4.6·T1 Td = 0.1·T1
PT3	**Tg/Tu**	**4.59**		
IAE	Kp·Ks = 5.4 Ti = 9.4·T1 Td = 0.7·T1	Kp·Ks = 7 Ti = 10·T1 Td = 0.7·T1	Kp·Ks = 8.4 Ti = 9.8·T1 Td = 0.7·T1	Kp·Ks = 10 Ti = 9.7·T1 Td = 0.7·T1
ITAE	Kp·Ks = 5.4 Ti = 9.4·T1 Td = 0.7·T1	Kp·Ks = 7 Ti = 10·T1 Td = 0.7·T1	Kp·Ks = 8.2 Ti = 9.6·T1 Td = 0.7·T1	Kp·Ks = 10 Ti = 9.7·T1 Td = 0.7·T1
ISE	Kp·Ks = 6.1 Ti = 10·T1 Td = 0.6·T1	Kp·Ks = 8.1 Ti = 9.8·T1 Td = 0.6·T1	Kp·Ks = 10 Ti = 10·T1 Td = 0.6·T1	Kp·Ks = 10 Ti = 7.8·T1 Td = 0.6·T1

PT4	**Tg/Tu**	**3.13**		
IAE	Kp·Ks = 2 Ti = 5.2·T1 Td = 1.1·T1	Kp·Ks = 2.9 Ti = 6.5·T1 Td = 1.2·T1	Kp·Ks = 3.3 Ti = 7.1·T1 Td = 1.3·T1	Kp·Ks = 3.3 Ti = 6.9·T1 Td = 1.3·T1
ITAE	Kp·Ks = 1.9 Ti = 5·T1 Td = 1.1·T1	Kp·Ks = 2.4 Ti = 5.9·T1 Td = 1.2·T1	Kp·Ks = 2.3 Ti = 5.7·T1 Td = 1.2·T1	Kp·Ks = 2.1 Ti = 5·T1 Td = 1.1·T1
ISE	Kp·Ks = 2.8 Ti = 6.6·T1 Td = 1.2·T1	Kp·Ks = 3.6 Ti = 7·T1 Td = 1.2·T1	Kp·Ks = 4.9 Ti = 7.1·T1 Td = 1.4·T1	Kp·Ks = 5.2 Ti = 7·T1 Td = 1.4·T1
PT5	**Tg/Tu**	**2.44**		
IAE	Kp·Ks = 1.7 Ti = 5.8·T1 Td = 1.6·T1	Kp·Ks = 1.8 Ti = 5.9·T1 Td = 1.6·T1	Kp·Ks = 1.8 Ti = 5.8·T1 Td = 1.6·T1	Kp·Ks = 1.7 Ti = 5.5·T1 Td = 1.6·T1
ITAE	Kp·Ks = 1.4 Ti = 5.3·T1 Td = 1.4·T1	Kp·Ks = 1.4 Ti = 5.2·T1 Td = 1.4·T1	Kp·Ks = 1.4 Ti = 5.2·T1 Td = 1.4·T1	Kp·Ks = 1.4 Ti = 5.0·T1 Td = 1.4·T1
ISE	Kp·Ks = 1.9 Ti = 5.9·T1 Td = 1.7·T1	Kp·Ks = 2.6 Ti = 6.5·T1 Td = 1.8·T1	Kp·Ks = 2.5 Ti = 6.3·T1 Td = 1.8·T1	Kp·Ks = 2.5 Ti = 6.1·T1 Td = 1.8·T1
PT6	**Tg/Tu:**	**2.03**		
IAE	Kp·Ks = 1.3 Ti = 5.9·T1 Td = 1.9·T1	Kp·Ks = 1.3 Ti = 5.8·T1 Td = 1.9·T1	Kp·Ks = 1.3 Ti = 5.8·T1 Td = 1.9·T1	Kp·Ks = 1.3 Ti = 5.6·T1 Td = 1.9·T1
ITAE	Kp·Ks = 1.1 Ti = 5.5·T1 Td =1.7·T1	Kp·Ks = 1.1 Ti = 5.5·T1 Td = 1.7·T1	Kp·Ks = 1.1 Ti = 5.4·T1 Td = 1.7·T1	Kp·Ks = 1.1 Ti = 5.3·T1 Td = 1.7·T1
ISE	Kp·Ks = 1.8 Ti = 6.8·T1 Td = 2.1·T1	Kp·Ks = 1.8 Ti = 6.5·T1 Td = 2.1·T1	Kp·Ks = 1.8 Ti = 6.5·T1 Td = 2.1·T1	Kp·Ks = 1.8 Ti = 6.3·T1 Td = 2.1·T1
PT8	**Tg/Tu:**	**1.56**		
IAE	Kp·Ks = 0.9 Ti = 6.3·T1 Td = 2.3·T1	Kp·Ks = 0.9 Ti = 6.3·T1 Td = 2.3·T1	Kp·Ks = 0.9 Ti = 6.2·T1 Td = 2.3·T1	Kp·Ks = 0.9 Ti = 6.1·T1 Td = 2.3·T1
ITAE	Kp·Ks = 0.8 Ti = 6·T1	Kp·Ks = 0.8 Ti = 5.9·T1	Kp·Ks = 0.8 Ti = 5.9·T1	Kp·Ks = 0.8 Ti = 5.8·T1

	Td =2.0·T1	Td = 2.0·T1	Td = 2.0·T1	Td = 2.0·T1
ISE	Kp·Ks = 1.2 Ti = 7·T1 Td = 2.7·T1	Kp·Ks = 1.2 Ti = 7·T1 Td = 2.7·T1	Kp·Ks = 1.2 Ti = 6.9·T1 Td = 2.7·T1	Kp·Ks = 1.2 Ti = 6.8·T1 Td = 2.7·T1
PT10	**Tg/Tu:**	**1.29**		
IAE	Kp·Ks = 0.8 Ti = 7.3·T1 Td = 2.8·T1	Kp·Ks = 0.8 Ti = 7.3·T1 Td = 2.8·T1	Kp·Ks = 0.8 Ti = 7.2·T1 Td = 2.8·T1	Kp·Ks = 0.8 Ti = 7.1·T1 Td = 2.8·T1
ITAE	Kp·Ks = 0.7 Ti = 6.9·T1 Td =2.4·T1	Kp·Ks = 0.7 Ti = 6.8·T1 Td = 2.4·T1	Kp·Ks = 0.7 Ti = 6.8·T1 Td = 2.4·T1	Kp·Ks = 0.7 Ti = 6.7·T1 Td = 2.4·T1
ISE	Kp·Ks = 1 Ti = 8·T1 Td = 3.2·T1	Kp·Ks = 1 Ti = 8·T1 Td = 3.2·T1	Kp·Ks = 1 Ti = 7.9·T1 Td = 3.2·T1	Kp·Ks = 1 Ti = 7.8·T1 Td = 3.2·T1
PT12	**Tg/Tu:**	**1.12**		
IAE	Kp·Ks = 0.7 Ti = 8.1·T1 Td = 3.1·T1	Kp·Ks = 0.7 Ti = 8.1·T1 Td = 3.1·T1	Kp·Ks = 0.7 Ti = 8.1·T1 Td = 3.1·T1	Kp·Ks = 0.7 Ti = 7.9·T1 Td = 3.1·T1
ITAE	Kp·Ks = 0.6 Ti = 7.5·T1 Td =2.6·T1	Kp·Ks = 0.6 Ti = 7.5·T1 Td = 2.6·T1	Kp·Ks = 0.6 Ti = 7.4·T1 Td = 2.6·T1	Kp·Ks = 0.6 Ti = 7.3·T1 Td = 2.5·T1
ISE	Kp·Ks = 0.9 Ti = 9.2·T1 Td = 3.6·T1	Kp·Ks = 0.9 Ti = 9.2·T1 Td = 3.6·T1	Kp·Ks = 0.9 Ti = 9.1·T1 Td = 3.7·T1	Kp·Ks = 0.9 Ti = 8.9·T1 Td = 3.7·T1
PT16	**Tg/Tu:**	**0.91**		
IAE	Kp·Ks = 0.6 Ti = 9.9·T1 Td = 3.9·T1	Kp·Ks = 0.6 Ti = 9.9·T1 Td = 3.8·T1	Kp·Ks = 0.6 Ti = 9.8·T1 Td = 3.9·T1	Kp·Ks = 0.6 Ti = 9.7·T1 Td = 3.8·T1
ITAE	Kp·Ks = 0.6 Ti = 9.9·T1 Td = 3.7·T1	Kp·Ks = 0.6 Ti = 9.9·T1 Td = 3.7·T1	Kp·Ks = 0.6 Ti = 9.8·T1 Td = 3.6·T1	Kp·Ks = 0.6 Ti = 9.7·T1 Td = 3.6·T1
ISE	Kp·Ks = 0.8 Ti =11.7·T1 Td = 4.5·T1	Kp·Ks = 0.8 Ti =11.6·T1 Td = 4.5·T1	Kp·Ks = 0.8 Ti =11.6·T1 Td = 4.5·T1	Kp·Ks = 0.8 Ti =11.4·T1 Td = 4.6·T1
PT20	**Tg/Tu:**	**0.77**		
	Kp·Ks = 0.6	Kp·Ks = 0.6	Kp·Ks = 0.6	Kp·Ks = 0.6

IAE	Ti =12.3·T1 Td = 4.8·T1	Ti =12.3·T1 Td = 4.8·T1	Ti =12.3·T1 Td = 4.8·T1	Ti =12.1·T1 Td = 4.8·T1
ITAE	Kp·Ks = 0.5 Ti =11.1·T1 Td =3.9·T1	Kp·Ks = 0.5 Ti =11.1·T1 Td = 3.9·T1	Kp·Ks = 0.5 Ti =11.1·T1 Td = 3.9·T1	Kp·Ks = 0.5 Ti = 11·T1 Td = 3.9·T1
ISE	Kp·Ks = 0.7 Ti =13.3·T1 Td = 5.5·T1	Kp·Ks = 0.7 Ti =13.3·T1 Td = 5.5·T1	Kp·Ks = 0.7 Ti =13.2·T1 Td = 5.6·T1	Kp·Ks = 0.7 Ti = 13·T1 Td = 5.7·T1

Table 3: Table values of the PID parameters for the minimum IAE, ITAE and ISE criteria of controlled PTn or time delayed systems.

4. Application of the table: Control of a PT2 system

For the first comparison, a didactic example is used to show the basic suitability of the parameter table. The response of a time-delayed system to a unit step shows a final static value of 1, a delay time Tu of 0.28 seconds and a rise time Tg of 2.72 seconds. This makes Tg/Tu = 9.65, resulting in PT2 behavior with Ks = 1 and T1 = 1 second.

For the ITAE criterion, the table values of the PID parameters for the PT2 system are read. With Ks = 1 and T1 = 1 for the regulation according to ITAE and the controller output limitation +/-2 is there: Kp = 10; Ti = 6.3; Td = 0.3. The simulation of the step responses of the closed loop systems according to figure 3 are shown in figure 5. They show a very nice transient response. The different dynamics or rise times can be explained well with the different controller output limitations.

This also shows very well that these must be included in the controller design. In the next example, a real problem is dealt with, with a much longer time delay, in order to also show the practicability of the presented parameter table.

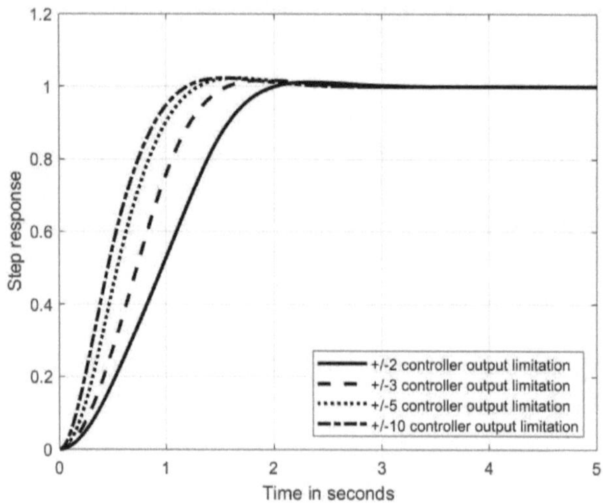

Figure 5: Step response of a closed loop system after the block diagram of figure 3 and the used table values after the ITAE criterion.

5. Application of the table: Control of a liquid concentration with dead time

In the system according to figure 1, two alcohol liquids with different flow rates Q_1 and Q_2 and alcohol concentrations c_1 and c_2 are mixed with one another using a stirrer. Another regulation not dealt with here ensures that the fill level in the mixer tank is always constant. In order to adjust the alcohol concentration c_0 of the liquid solutions after the stirring process, the concentration c_1 of the first liquid solution can be changed, while the second solution can not be changed.

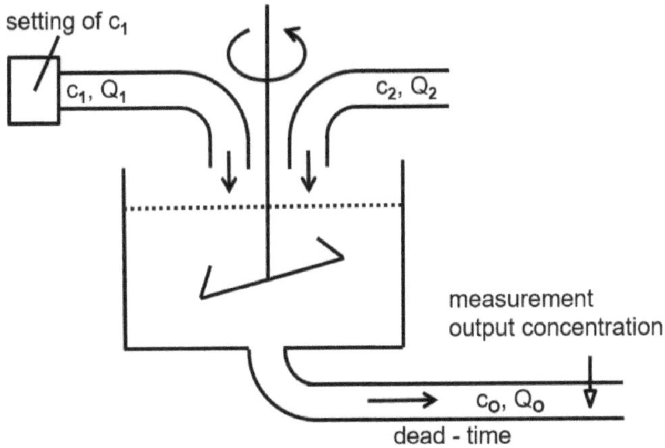

Figure 6: Technological scheme of a system to adjust the alcohol concentration c_O in a liquid.

The following formulas are therefore valid in the stationary case:

$$Q_1 + Q_2 = Q_O \qquad and \qquad Q_1 \cdot c_1 + Q_2 \cdot c_2 = Q_O \cdot c_O$$

The system has the following data at the operating point: Liquid 1; Q_1 = 2l/s and c_1 = 20%, Liquid 2; Q_2 = 1l/s and c_2 = 60%. This results in a flow rate of 3l/s for Q_O and the concentration c_O is calculated as 33.33% using the formula below

$$c_O = \frac{Q_1 \cdot c_1 + Q_2 \cdot c_2}{Q_O}$$

One would now like to regulate the concentration of c_O after the mixture with a PID controller. The controller output is the concentration change c_1 and the controlled variable is c_O. The concentration is measured according to figure 6 in the flow tube after the mixer. Since the liquid has a constant volume flow after mixing and has to cover a distance in the pipe to

reach the measuring point, there is also a dead time. The mixing process itself also has dynamics and this results in the classic and well-known behavior of a system with dead time.
In order to be able to calculate the PID controller with the parameter sets from the table, firstly the step response of the above system must be measured.
For this purpose, the concentration c_1 is increased from 20% to 40% at time t=0. Another control circuit is also responsible for this in the system, but it is very fast in comparison and is therefore not taken into account in the system dynamics.
The concentration c_O is calculated according to the above formula in the stationary case to be 46.33%.
Figure 7 shows the measurement of the step response of c_O for a step of c_1 from 20% to 40%.
Where Ks = difference of the output, Δc_O divided by delta of the input, Δc_1, compared after and before the step. In this specific case, this also results in:

$$K_s = \frac{\Delta c_O}{\Delta c_1} = \frac{46.66\% - 33.33\%}{40\% - 20\%} = 0.67$$

Figure 7 shows that there is a very large dead time Tu of 43 seconds. The subsequent rise time Tg is 39 seconds. The ratio Tg/Tu is therefore 39/43 = 0.91. According to the table values of the turning tangent method, this corresponds fairly well to a PT16 behavior, i.e. the behavior of 16 PT1 elements connected in series. Furthermore, the identical time constants are T1 = Tg/9.76 = 39/9.76. That gives about T1 = 4 seconds.

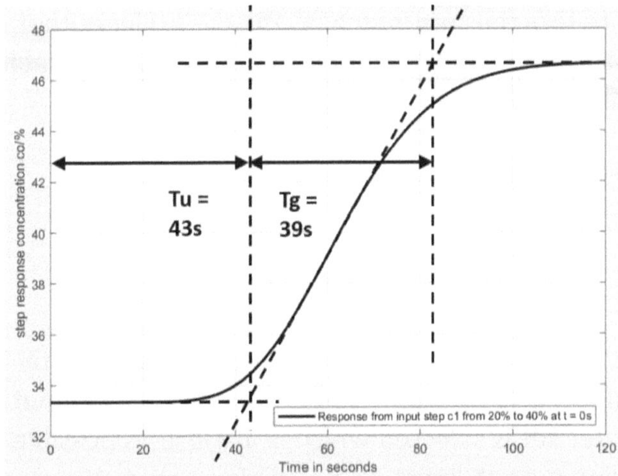

Figure 7: Step response of the controlled variable c_O after a step of c_1 from 20% to 40%.

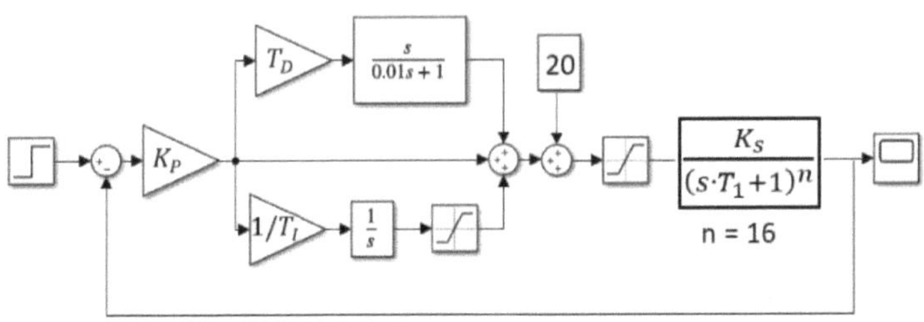

Figure 8: Block diagram of the closed loop system, including the operating point.

In order to determine the table values of the optimal controller parameters according to the IAE or ITAE criterion, the controller output limitation is also required. As an example, you want to achieve a step in the initial concentration from 33.33% to 46.67%. In this system, the concentration c_1 can be a maximum of 60%. The controller output before the step was 20%. The controller output for the stationary end value is 40. According to the table, the controller output limitation is:

$$\text{Control output limitation} = \frac{\text{Maximum controller output} - \text{controller output before the step}}{\text{controller output for the stationary end value} - \text{controller output before the step}} = \frac{60\% - 20\%}{40\% - 20\%} = 2$$

It is of course clear that if different steps are to be implemented with the same controller parameters, a parameter set is selected which is as close as possible to the controller output limits to be implemented. However, the parameter sets in the table for the individual controller output limitations are very similar. So, this is not a major limitation.

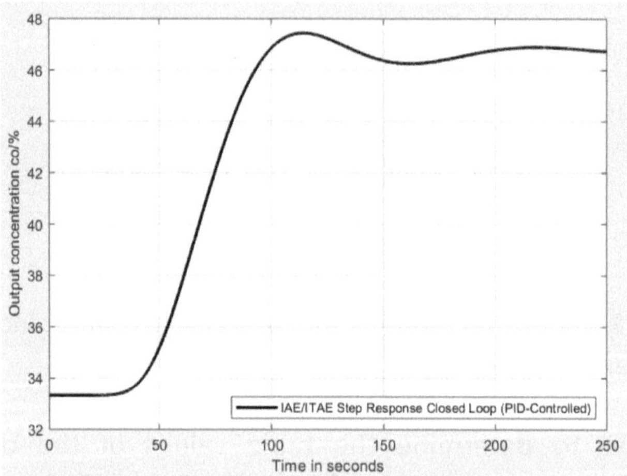

Figure 9: Step response of the closed loop system, a step of c_O from 33.33% to 46.67%.

In the table one finds almost the same values for the two criteria IAE and ITAE PT16, Tg/Tu = 0.91 and for a controller output limitation of +/-2 for below applies to ITAE:

$$Kp = \frac{0.6}{Ks} = 0.9 \qquad Ti = 9.9 \cdot T_1 = 36.6s \qquad Td = 3.7 \cdot T_1 = 14.8s$$

If one chooses the controller structure according to figure 8 and usees the above values as parameters, one gets get the curve according to figure 9 as the step response of the closed loop.

6. Discussion

The examples discussed show excellent transient response. However, the table can be used in general. It is not comparable to heuristic methods, because the parameters in the table are precise calculated values that minimize the quality criteria. It is also up for discussion, what would happen if different steps were executed and the parameters therefore had to be selected according to different controller output limitation factors. However, it turns out that the parameters are very similar and you should choose values for steps that are most likely to occur in the specific system. In the case of non-linear systems, the Ks and T1 are also different if the systems have different step responses. There you have to carry out more precise analyzes of the meaningful parameters.

An exciting insight comes from the discussion of the question of how the parameters Kp, Ti and Td develop for changing ratios Tg/Tu (i.e. rise time in relation to delay time).

The smaller the ratio becomes, i.e. the larger the delay time becomes in relation to the rise time, the smaller Kp becomes on the one hand and the larger Td becomes on the other.

The effect of the small Kp, in the example according to the table the multiplication of Kp by Ks is only 0.6 means that the system can only be controlled slowly. In the literature, this is also described in such a way that the controllability is reduced for systems with longer dead times. If you were to plot the controller output as well, you would see that this is also relatively small. So it is of no use with these systems if you provide an additional controller reserve through amplifiers, because this cannot be used at all due to the system. This also means that the controlled system does not behave any more dynamically than the uncontrolled one, as a comparison of the step responses in the figures 7 and 9 shows.

However, a control should achieve two basic things: on the one hand, one often wants to make the system faster and provides the system with greater gains and thus a higher controller output in the short term, and on the other hand, one wants the measurement of the controlled variable, in this example the output concentration c_O, follows the target value exactly, especially with the static final value.

If one follows the development of the value of Kp in the tables, then larger system dynamics are no longer achieved with smaller ratios Tg/Tu (or larger orders n of the PTn systems and thus larger dead times) and smaller Kp. On the one hand, however, the controlled variable of the controlled system shows a nice transient behavior according to the minimized quality criteria and, in particular, also reaches the setpoint in the steady-state final value. In practice, this is often sufficient for such dead-time systems

When looking at the differential component of the controller Td, it becomes clear that there is no differential component for the optimization of the quality criteria when controlling frequently occurring PT1 elements, i.e. a pure PI controller is

already optimal. As the system order increases, i.e. the ratio Tg/Tu decreases, or the dead time Tu increases, the required D component (Td) increases.
In practice, one can often do without a simulation and only measure the step response of the system. Then the order n and the associated parameters for the PID controller can be read from the table and the controller can be implemented directly in the system.
This booklet aims to make a contribution to greatly simplify the development process of controller design for such in practical work frequently occurring systems.

7. Literature

[1] Ziegler, J. B., Nichols N. B., Optimum settings for automatic controllers, ASME Transactions, v64 (1942), pp. 759-768

[2] Kun Li Chien Kun Li, Hrones, J. A., Reswick, J. B., On the Automatic Control of Generalized Passive Systems. In: Transactions of the American Society of Mechanical Engineers., Bd. 74, Cambridge (Mass.), USA, Feb. 1952, S. 175–185

[3] Russel, Stuart J, Norvig, Peter, Artificial Intelligence: A Modern Approach (2nd edition), Upper Saddle River, New Jersey: Prentice Hall, pp 111- 114, ISBN 0-13-790395-2, 2003

[4] Büchi, Roland. "PID Controller Parameter Tables for Time-Delayed Systems Optimized Using Hill-Climbing." *Signals* 3.1 (2022): 146-156.

[5] da Silva LR, Flesch RC, Normey-Rico JE. Controlling industrial dead-time systems: When to use a PID or an advanced controller. ISA transactions. 2020 Apr 1;99:339-50.

[6] Qi, Zhi, Qian Shi, and Hui Zhang. "Tuning of digital PID controllers using particle swarm optimization algorithm for a CAN-based DC motor subject to stochastic delays." IEEE Transactions on Industrial Electronics 67.7 (2019): 5637-5646.

[7] Joseph, E. A., Olaiya, O. O., Cohen- Coon PID Tuning Method, A Better Option to Ziegler Nichols- PID Tuning Method, Computer Engineering and Intelligent Systems, ISSN 2222-1719, Vol. 9, No. 5, 2018

[8] Unbehauen H. Regelungstechnik. Braunschweig: Vieweg; 1992.

[9] S. Zacher, M. Reuter: Regelungstechnik für Ingenieure. 15.Auflage, Springer Vieweg Verlag, 2017

[10] Büchi, Roland. "Brushless-Motoren und–Regler. 1." *Auflage, Baden-Baden: Verlag für Technik und Handwerk neue Medien GmbH* (2013).

[11] G. Schwarze: Bestimmung der regelunsgtechnischen Kennwerte von P-Gliedern aus der Übergangsfunktion ohne Wendetangentenkonstruktion, In: - messen-steuern-regeln Heft 5, S. 447-449, 1962

[12] Martins, Fernando G., Tuning PID Controllers using the ITAE Criterion, Int. Journal Engineering, Edition. Vol. 21, No. 5. Pp. 867-873, 2005

[13] Büchi, Roland. *State space control, LQR and observer: step by step introduction with Matlab examples.* Norderstedt Books on Demand, 2010.

[14] Leishman Guillermo J. Silva, Aniruddha Datta. S. P. Bhattacharyya. “PID Controllers for Time-Delay Systems”. Boston. ISBN 0-8176-4266- 8.2005.